Health
238

来自太空的病毒？

Virus from Outer Space?

Gunter Pauli

[比] 冈特·鲍利 著

[哥伦] 凯瑟琳娜·巴赫 绘

章里西 译

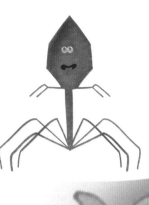

上海远东出版社

丛书编委会

主　任：贾　峰

副主任：何家振　闫世东　郑立明

委　员：李原原　祝真旭　牛玲娟　梁雅丽　任泽林

　　　　王　岢　陈　卫　郑循如　吴建民　彭　勇

　　　　王梦雨　戴　虹　靳增江　孟　蝶　崔晓晓

特别感谢以下热心人士对童书工作的支持：

匡志强　方　芳　宋小华　解　东　厉　云　李　婧

刘　丹　熊彩虹　罗淑怡　旷　婉　杨　荣　刘学振

何圣霖　王必斗　潘林平　熊志强　廖清州　谭燕宁

王　征　白　纯　张林霞　寿颖慧　罗　佳　傅　俊

胡海朋　白永喆　韦小宏　李　杰　欧　亮

目录

Contents

一棵藜麦盯着头上的天空出了神。一头母牛在边上看到他如痴如醉的样子，说道：

"耐心点，我的朋友，在晚上这个时候看流星也太早了吧。"

"哦，我并不是想看星星。我在看上头到底会掉下来多少病毒。"

$\mathscr{A}$ quinoa plant is staring up at the sky. A cow is watching him and, seeing his dreamy expression, she says:

"$\mathscr{Y}$ou'll have to be patient, my friend. It is too early in the evening to see any falling stars."

"$\mathscr{O}$h, it's not stars I want to see. I am checking to see how many viruses are falling down on me."

一棵藜麦盯着头上的天空出了神。

A quinoa plant is staring up at the sky.

5

......病毒对我们的生存其实是必需的。

... viruses are very much needed.

"哦！别提病毒！我可不想被这些东西感染。几百万只非洲的水牛、长颈鹿和角马等偶蹄动物因病毒感染丧生，结果是我们牛群被扣上了传播病毒到那里的罪名。"

"好吧，不管你喜不喜欢，病毒对我们的生存其实是必需的。"

"Oh no! Not viruses! I don't want to be infected by any of those. We cattle have already been blamed for spreading a viral disease to Africa, where it killed millions of cloven-hoofed animals, like buffalo, giraffe and wildebeest."

"Well, whether you like it or not, viruses are very much needed."

"必需？不可能！病毒会让你生病，甚至会让你丧命。"

"可是到目前为止，它们确实是地球上最丰富的生命形式！"

"这真是非常吓人的消息。我原以为病毒根本就不算生命……"

"Needed? No ways! Viruses make you sick, and can even kill you."

"Yet, they are the most abundant life form on the planet – by far!"

"That's very scary news, indeed. And I thought that viruses weren't even alive ..."

病毒会让你生病……

Viruses make you sick...

9

……这绝不意味着它就不是生命。

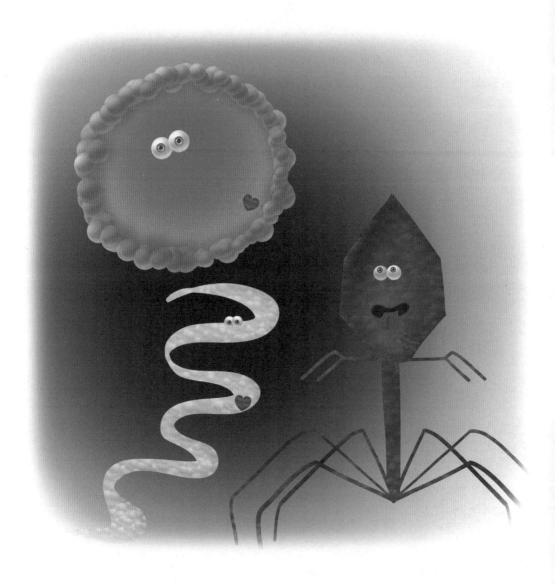

... doesn't mean it is not considered a living thing.

"病毒的确不能像动植物那样繁殖，但这绝不意味着它就不是生命。"

"我听说病毒必须侵入并最终占领寄主才能生存下去。"

"Well, just because a virus doesn't reproduce like an animal or a plant, it doesn't mean that it is not considered a living thing."
"I heard that a virus must invade, and take over, to live on."

"没错，它们捕食快速生长的细菌和藻类。但你要知道，并非所有的病毒都是有害的，有些病毒甚至是有益的。我觉得人们还不理解病毒对我们的重要性。"

　　"地球上最庞大的捕食种群？对我们很重要？你想让我相信病毒有什么好处，这可不太容易。"

"They hunt bacteria and even algae that are blooming. All viruses aren't bad, you know. Some do good, and I don't think we have even started understanding how important they can be for us."

"The greatest predators on Earth?Important for us? You'll have a hard time convincing me that viruses are any good."

它们捕食细菌……

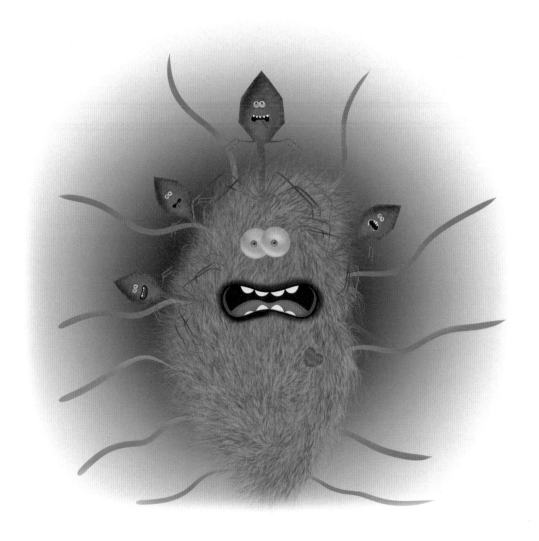

They hunt bacteria …

......吸收二氧化碳并且释放氧气。

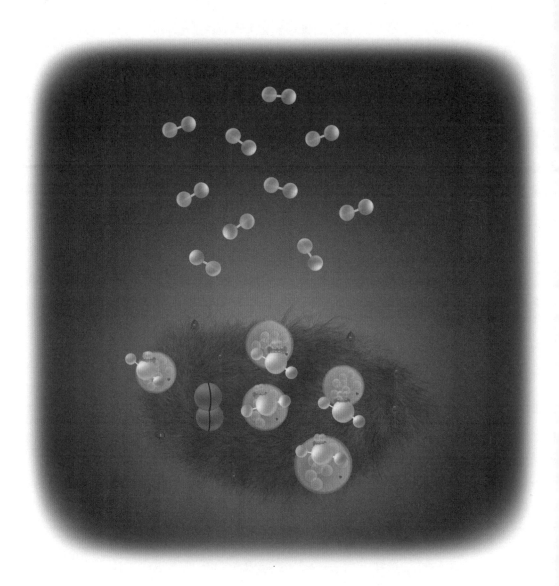

... eat carbon dioxide and produce oxygen.

"你看，病毒可以释放食物。就像麋鹿被一群灰狼杀死后成为郊狼和渡鸦的食物一样，被病毒杀死的微生物也是浮游生物的食物，浮游生物吸收二氧化碳并且释放氧气。"

"我知道我们需要氧气来呼吸，但这是病毒唯一的好处吗？"

"Look, viruses free up food. In the same way that an elk, killed by a pack of wolves, becomes food for coyotes and ravens, a microbe killed by a virus is food for plankton that eat carbon dioxide and produce oxygen."

"I know we need oxygen to breathe, but is that the only good thing viruses do?"

"就我个人而言，如果没有它们，我不可能像在高地上那样的地方生长和生产粮食。是病毒帮我度过每一次干旱。我的朋友真菌也需要一种病毒才能健康成长。"

"保持健康？但是藜麦，要知道讨厌的病毒会让你生重病的。"

"但它们是我们生活的一部分。"

"I, for one, cannot possibly grow and produce food as well as I do on the highlands without them. They help me survive every drought. And my friend, the fungus, needs a virus too, to grow and be healthy."

"Be healthy? But Quinoa, you know that nasty viruses can make you very sick."

"But they are part of our lives."

……我的朋友真菌也需要一种病毒才能健康成长。

... my friend, the fungus, needs a virus too.

... people have invented vaccines ...

"幸运的是，我们牛学会了保护自己不受病毒感染，而人们也发明了消灭病毒感染的疫苗。我一直想知道这些看不见的病毒是从哪里来的。"

"大多数病毒被海浪卷出水面，浮着在灰尘颗粒上，被风或者风暴推向高空。"

"看到浪花和风暴了吗？它们是怎么飞到世界各地的？"

"Fortunately, we cows have learnt to protect ourselves against them, and people have invented vaccines that wipe out viral infections. You know, I've always wondered where these invisible viruses come from."

"Most of them are swept out of the water by sea spray, or float on dust particles pushed high in the sky by winds and storms."

"See spray and storms, really? And how do they jet around the globe?"

"哦，病毒变成了旅行家，通过急流环绕地球，然后又落到地球的每个角落。"

"席卷非洲，然后降落在中国。这听起来很疯狂。"

"嗯，一些科学家甚至相信病毒来自外太空！"

……这仅仅是开始！……

"Oh, viruses turn into globetrotters circling the planet by jet streams, and then they fall back down, anywhere on Earth."

"Swept up in Africa, and deposited in China. That sounds crazy."

"Well, some scientists even believe that viruses come from other planets!"

... AND IT HAS ONLY JUST BEGUN!...

... AND IT HAS ONLY JUST BEGUN! ...

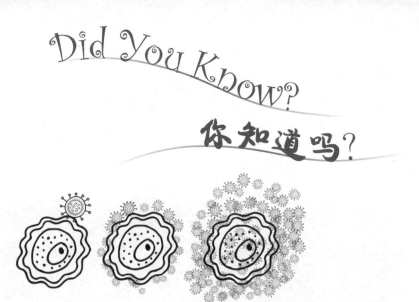

When infected by a flu virus, one cell is able to produce 10,000 new viruses, increasing to 100 trillion, in a matter of days. This dynamic multiplication explains why there are 10 million times more viruses on Earth than there are stars in the universe.

当被流感病毒感染时，一个细胞能产生1万个新的病毒，短短几天病毒数量就会增加到100万亿个。这种动态增殖解释了为什么地球上的病毒比宇宙中的恒星要多1000万倍。

It has been estimated that approximately 800 million viruses cascade into every square metre of the planet every day. Our immune systems must function well to handle this onslaught of potential infections.

据估计，每天大约有8亿个病毒落在地球上的每平方米土地上。我们的免疫系统必须运转良好，才能应对潜在感染的冲击。

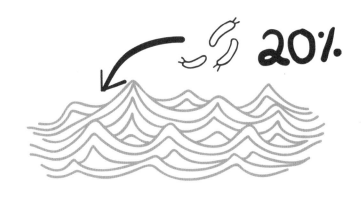

Viruses cause trillions and trillions of infections in the ocean, causing the destruction of 20 percent of all bacterial cells in the sea every day, where they outnumber bacteria by a factor of 10 to one.

病毒每天在海洋中发生几万亿次传播，能够杀死海洋中 20% 的细菌，它们的数量是细菌的 10 倍。

An ancient virus inserted its DNA into the genomes of four-limbed animals that were ancestors to humans. That tiny piece of DNA is part of human nervous system and plays a role in nerves and memory.

一种古老的病毒将其 DNA 植入人类祖先四肢动物的基因组中。这一小段 DNA 是人类神经系统的一部分，在神经和记忆中发挥作用。

Viruses keep ecosystems in balance. Viruses attack toxic algae blooms in the ocean, ending an outbreak in a few days, and converting these to food for other ocean dwellers.

病毒维持着生态系统的平衡。病毒攻击海洋中大量繁殖的有毒藻类，能够在几天内终止藻类爆发，并将它们转化为其他海洋生物的食物。

Some researchers claim that 8% (and others that 40-80%) of the human genome may be linked to ancient viral invasions that left their DNA behind. We have discovered the symbiotic relationship with bacteria, but still have to understand the role of viruses.

一些研究人员声称，8%（还有的说是 40%—80%）的人类基因组可能与留下 DNA 的远古病毒入侵有关。我们已经发现了与细菌的共生关系，但仍然需要了解病毒的作用。

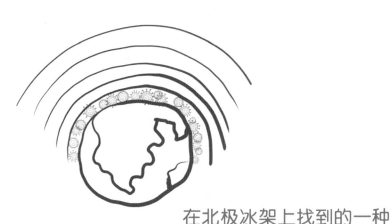

在北极冰架上找到的一种病毒和在墨西哥湾与德国找到的一种病毒一模一样。这些病毒可以随着海上溅起的水雾和地面飘起的尘埃进入大气对流层，在气候条件合适时从空中降落到地球表面。

A virus in the Arctic ice shelf is identical to a virus in the Gulf of Mexico, and to one in Germany. These viruses are transported via sea spray and soil dust into the troposphere and descend to the Earth's surface when climate conditions are right.

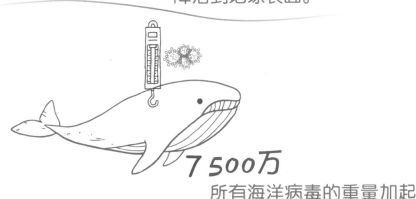

7 500万

所有海洋病毒的重量加起来超过 7 500 万只蓝鲸。如果你把所有海洋病毒一个个排成一条直线（每个 100 纳米），它们将延伸到最近的 60 个星系之外。

The combined weight of all ocean viruses exceeds the weight of 75 million blue whales. If you were to place all marine viruses end-to-end in a straight line (100 nanometres each), they would stretch beyond the nearest 60 galaxies.

There are many times more viruses than bacteria on Earth. Is this good or bad news?

地球上的病毒比细菌多很多倍，这是好消息还是坏消息？

Do you think that viruses come from Outer Space?

你认为病毒来自外太空吗？

Would you prefer an attack by a virus, or one by bacteria?

你喜欢被病毒攻击还是被细菌攻击？

Are viruses good for something?

病毒有什么好处吗？

What is the difference between a virus and a bacterium? And which illnesses are caused by viruses, and which by bacteria? An antibiotic (against bacteria) kills bacteria, but you need an antiviral to control a virus. To ensure that we receive the right medicine when ill, we need to know the difference. Ask your friends and family members if they know the difference, and how to treat the cause and the symptoms. Now ask a medical doctor how he or she goes about establishing whether a patient has a viral or a bacterial infection. When you have a clear idea of the difference, summarise your findings and share this valuable information with the people close to you.

病毒和细菌有什么区别？哪些疾病是病毒引起的，哪些是细菌引起的？抗生素可以杀死细菌，但你需要一种抗病毒药物来控制病毒。为了生病时吃对药，我们需要知道二者的区别。问问亲朋好友是否知道这种区别，知不知道如何来判断。问问医生是怎么诊断病毒感染和细菌感染的。搞清楚其中的区别后做个总结，并分享给亲近的人。

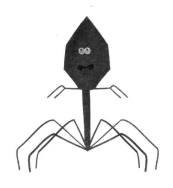

学科知识
Academic Knowledge

生物学	病毒是最简单最小的生命形式，比细菌小10到100倍；病毒通过入侵宿主的分子进行自我复制；病毒样的元素促进了DNA的进化，首批细胞的形成；生命形式：古生菌、细菌和真核生物；超大型病毒：潘多拉病毒和脊髓灰质炎病毒；天花是一种大DNA病毒；RNA病毒感染人类、植物、真菌和原生动物（真核生物）；感染古生菌、细菌和真核生物的病毒有几种相同的蛋白质。
化学	逆转录酶将RNA转化为DNA；革兰氏阴性菌对抗生素具有高耐药性；青霉素只能治疗细菌感染，不能治疗病毒感染；病毒基因组中核苷酸序列的改变；病毒含有核酸，要么是DNA（脱氧核糖核酸）要么是RNA（核糖核酸）；大蒜含有蒜素和大蒜烯，可以杀死病毒。
物理	病毒以无与伦比的速度复制；藜麦通过改变光合作用、呼吸作用、水分关系以及抗氧化和激素代谢来应对干旱，这些通常由病毒触发。
工程学	重组DNA方法是对病毒进行基因工程，将外源基因引入细胞，用于生物医学、农业、生物防治或技术目的；基因治疗，用于癌症治疗和疫苗的病毒基因工程；带有基因工程病毒的锂离子电池将用于阴极。
经济学	病毒具有积极和消极的经济重要性；病毒被用于制备疫苗、控制有害昆虫、控制疾病以及作为有机纳米颗粒；计算机病毒的出现；艾滋病毒、埃博拉病毒、天花和小儿麻痹症等病毒感染对社会的代价。
伦理学	"活着"的含义：有能力复制基因；一种巨型病毒或它的祖先先于其他类型的生命存在，并可能在我们所知的生命形成过程中扮演重要角色，这迫使我们重新思考生命起源理论。
历史	病毒世界假说和RNA世界假说；人类起源于非洲，非洲居民的基因多样性比其他任何地方都要多；牛瘟是1887年由印度和欧洲引入北非的一种病毒性疾病，导致非洲各地几百万只偶蹄家畜和野生动物死亡。
地理	病毒从地面被席卷到天气系统和平流层之间的大气层；细菌和病毒通常是通过雨水和撒哈拉沙漠尘埃而落回地球。
数学	海洋中有10^{31}个病毒颗粒；病毒数量是通过外推法计算的。
生活方式	症状可能相同，但病因不同；由细菌引起的肺炎、脑膜炎、链球菌性喉炎、耳部感染、伤口感染、淋病，由大肠杆菌引起的食物中毒；病毒会引起流感、水痘、普通感冒、乙型肝炎、风疹、SARS、麻疹、埃博拉、HPV、疱疹、狂犬病、艾滋病。
社会学	病毒被认为是一种有毒液体。
心理学	由于人们的贪婪、好奇和恐惧，电脑黑客能够成功地部署病毒；免疫系统、神经系统和心理系统紧密交织在一起。
系统论	病毒为细胞的发育提供原材料，继而入侵生命的每个角落。

情感智慧
Emotional Intelligence

牛

牛请藜麦保持耐心，同时也不理解藜麦对病毒的兴趣。她担心病毒有害身体健康。当听说病毒是生命的重要组成部分时，她拒绝接受这个事实，坚持认为病毒是有害的。她承认病毒在提供氧气方面间接发挥了作用。她认为病毒是一种入侵并控制生命的非生物，并宣称她还没有准备好相信病毒的好处。她自豪地宣布牛能够保护自己免受病毒的侵害，知道人类已经研制出疫苗。她对病毒是从哪里来的很感兴趣，不太理解藜麦的解释。即使有事实摆在面前，她也觉得难以置信。

藜麦

藜麦公开表示他对病毒有兴趣，并自信地表达出一种截然不同的观点，解释病毒对生命至关重要。他知道病毒具有多样性。他耐心地解释说，即使病毒不繁殖，它们也可以被认为是一种生命形式。他坚持认为病毒也是有好处的，当牛提出质疑时，他做了更详细的说明。他通过对比帮助牛来理解。他分享了自己在没有病毒存在的情况下无法生存的信息。尽管牛不相信，他还是耐心地解释病毒是从哪里来的，以及病毒如何在地球的每个角落繁衍。当牛对于病毒的疯狂传播而感到吃惊时，藜麦转移了话题，说一些科学家甚至相信病毒是来自外太空的。

艺术
The Arts

你是否从病毒的故事中得到了灵感？是不是吃惊它有令人难以置信的球形形状？这种形状称为二十面体。选用任何材料来创造一个二十面体。核心结构是一个由三角形组合而成的球体。你会发现科学和艺术是如何相融的，更重要的是，你可以借助艺术来理解病毒。

思维拓展
Systems: Making the Connections

无论我们走到哪里，都被独特的细菌和病毒所包围。所有的事物都在不断地共同进化，病毒深刻地塑造了生命和生物世界。病毒不属于生物，但它们与地球上的所有生物都有着长久的共同关系。病毒是一种强大的、具有调节作用的存在，特别是在海洋中，它被认为是影响全球气候的重要因素。如果没有病毒，我们今天所知道和依赖的自然世界就不会也不可能存在。如果没有病毒，地球甚至可能永远不会有生命。大约40亿年前，病毒可能是首批细胞的前身，充当着非生命和生命之间的重要桥梁。病毒每24小时杀死大约20%的海洋细菌。海洋中的病毒感染率令人难以置信。海洋每时每刻都在被病毒改变。迄今为止，科学家已经识别并命名了大约5 000种不同的病毒，但仍有数百万种病毒有待发现。如果我们无视病毒的存在，仅仅把它们视为"杀手"，我们对生命和可持续性又能了解多少呢？在人类基因组中发现的5%到8%的遗传物质来自病毒，一些人声称细菌成分最初也来自病毒。有证据清楚地表明，在漫长的岁月里，病毒一直是我们最初的、不变的伙伴。我们对病毒知道得越多，对病毒和我们自己的认识就会越深。

动手能力
Capacity to Implement

一种最常见的病毒感染是疱疹，由单纯疱疹病毒引起。它通常通过密切接触来传播，比如共用牙刷。在疱疹爆发后，这种病毒会退回到神经系统，保持隐身状态。因此，一旦组织被感染，通常没有永久的治愈方法。然而，也有一些天然的治疗手段，比如苹果醋具有收敛性、抗炎和抗氧化的作用。过氧化氢也是一种很好的天然消毒剂，尽管不一定杀死所有的病毒。经常吃大蒜也可以抑制病毒，但你不得不接受难闻的口气。

故事灵感来自
This Fable Is Inspired by

伊莎贝尔·雷切
Isabel Reche

伊莎贝尔·雷切现居西班牙，1995年她在格拉纳达大学获得博士学位。为此，她研究了高山湖中营养物质循环利用对微生物食物网的影响。1995年到1998年，她在美国纽约卡里生态系统研究学院担任博士后研究员。这是一个独立的非营利性环境研究组织，致力于世界生态系统的科学研究，以及影响生态系统的自然和人为因素研究。正是这个研究学院首次对酸雨现象进行了研究。伊莎贝尔研究撒哈拉沙尘对水生生态系统的影响，以及沙尘作为微生物和病毒传播媒介的作用。伊莎贝尔现在是格拉纳达大学生态学系教授，教授微生物地理学，以及"科学中的创造力、严谨性和交流"课程。

图书在版编目（CIP）数据

冈特生态童书. 第七辑：全36册：汉英对照 /
（比）冈特·鲍利著；（哥伦）凯瑟琳娜·巴赫绘；
何家振等译. —上海：上海远东出版社，2020
ISBN 978-7-5476-1671-0

Ⅰ.①冈… Ⅱ.①冈… ②凯… ③何… Ⅲ.①生态
环境–环境保护–儿童读物—汉英 Ⅳ.①X171.1-49

中国版本图书馆CIP数据核字（2020）第236911号

策　　划　张　蓉
责任编辑　程云琦
封面设计　魏　来李　廉

冈特生态童书
来自太空的病毒？
[比]冈特·鲍利　著
[哥伦]凯瑟琳娜·巴赫　绘
章里西　译

记得要和身边的小朋友分享环保知识哦！
八喜冰淇淋祝你成为环保小使者！